SpringerBriefs in Applied Sciences and Technology

SpringerBriefs in Computational Intelligence

Series Editor

Janusz Kacprzyk, Systems Research Institute, Polish Academy of Sciences, Warsaw, Poland

SpringerBriefs in Computational Intelligence are a series of slim high-quality publications encompassing the entire spectrum of Computational Intelligence. Featuring compact volumes of 50 to 125 pages (approximately 20,000-45,000 words), Briefs are shorter than a conventional book but longer than a journal article. Thus Briefs serve as timely, concise tools for students, researchers, and professionals.

More information about this subseries at http://www.springer.com/series/10618

Nguyen Thi Ngoc Anh · Tran Ngoc Thang ·
Vijender Kumar Solanki

Artificial Intelligence for Automated Pricing Based on Product Descriptions

 Springer

Nguyen Thi Ngoc Anh
School of Applied Mathematics
and Informatics
Hanoi University of Science
and Technology
Hanoi, Vietnam

Tran Ngoc Thang
School of Applied Mathematics
and Informatics
Hanoi University of Science
and Technology
Hanoi, Vietnam

Vijender Kumar Solanki
CMR Institute of Technology
(Autonomous)
Hyderabad, Telangana, India

ISSN 2191-530X ISSN 2191-5318 (electronic)
SpringerBriefs in Applied Sciences and Technology
ISSN 2625-3704 ISSN 2625-3712 (electronic)
SpringerBriefs in Computational Intelligence
ISBN 978-981-16-4701-7 ISBN 978-981-16-4702-4 (eBook)
https://doi.org/10.1007/978-981-16-4702-4

This Springer imprint is published by the registered company Springer Nature Singapore Pte Ltd.
The registered company address is: 152 Beach Road, #21-01/04 Gateway East, Singapore 189721, Singapore

We think it is important in the science of data analysis to maintain the joy of finding the laws of data. We started to feel as if we were truly responsible for machine learning's success in data analysis. Discovering the laws of data always feels interesting. The world has and will have a lot of data; data analysis will help people understand more deeply about how companies are operating. The key to a successful machine learning paradigm is in our hands. With the contribution in this book, we wish to contribute to the development of the science of data analysis.

Preface I

Over the past decades, as data are produced and collected increasingly large, researchers need to apply suitable methods to analyze this data. Statistical methods and machine learning were used to analyze this volume of data.

In our experience, application researchers still find it difficult to spend extra time evaluating data in product pricing. This book is developed for researchers looking for a solution to the problem of analyzing data from product descriptions to product prices. The book was originally developed for two published papers, but we feel that applied researchers from other disciplines may find this book useful as well. Furthermore, we actively advise on models and techniques used in product pricing. These practical guidelines can be useful to help researchers get started with the desire to price products from the data.

With this book, we provide an overview of the available methods for pricing products. The use of the methods is well explained and supported by practical examples. We hope you will enjoy this book and you find it helpful, as a result you will use recommended methods to solve product pricing problems from your data.

Hanoi, Vietnam   Nguyen Thi Ngoc Anh
April 2021   Tran Ngoc Thang
                                             Vijender Kumar Solanki

Preface II

Data analysis is written for data analysts looking to find data mining. In writing this book, we recognize that there is a lot of literature for product pricing primarily derived from supply-demand relationships, economic policy and partly from data analysis. We have tried to present a valuation of data analysis, an area of technical topics but applied in the economic realm.

We started the book by introducing the pricing problem from product descriptions, methods used in pricing. In Chap. 2, we described how to use NLP in extracting data from product descriptions for pricing. In Chap. 3, we have grouped and turned qualitative attributes into quantitative for more accurate pricing. In Chap. 4, we present machine learning methods, essebmle methods, and GA algorithm to optimize the architecture of machine learning. In the final chapter, we give some application and discussion.

We welcome feedback. If you have an idea of how we can make this book better—or topics you'd like to cover in a new edition, we'd love to hear from you. Please contact us via email: anh.nguyenthingoc@hust.edu.vn, thang.tranngoc@hust.edu.vn, spesinfo@yahoo.com.

Hanoi, Vietnam
April 2021

Nguyen Thi Ngoc Anh
Tran Ngoc Thang
Vijender Kumar Solanki

Acknowledgements

Writing a book is harder than we thought but more useful than we could have imagined. None of us would have been able to do it without our colleague who has always supported and helped.

Writing a book on product pricing has a lot of editorial help, insight, and relentless support in bringing our story to life. Thanks to their efforts and encouragement, we have a complete book to recommend to everyone.

For our family. A very special thanks to our family members, who have always accompanied and supported us in life.

Finally, to all those who have participated in our work: Vu Thanh Nam, Dang Minh Tuan, Pham Ngoc Linh, Nguyen Thanh Binh, Nguyen Huy Anh, Nguyen Nhat Anh, Doan Van Thai, Luong Ngoc Son, Pham Vu Tien, Luong Khac Manh.

Hanoi, Vietnam
April 2021

Nguyen Thi Ngoc Anh
Tran Ngoc Thang
Vijender Kumar Solanki

Contents

Acronyms

AI	Artificial Intelligent
ANOVA	Analysis of variance
GA	Genetic algorithm
LSTM	Long-short term memory
ML	Machine learning
NLP	Nature language processing

Chapter 1
Pricing Based on Product Descriptions: Problem, Data and Methods

1.1 Pricing Problems

1.1.1 Pricing Products

The problem of determining the price of goods based on the description of the goods is the problem of automatically giving information and the value of the goods corresponding to the description of the goods. The input to the problem is a description of the good, the product, and the output will be the price associated with that description. In the manual process, the employee will have to read the declaration and query the price database in order to determine the price of the goods. Here, the price database can understand it as a price list corresponding to the specific attributes of the product and for a specific period of time there will also be different prices. However, because the number of declarations is very large and the content of the declaration often has many types of descriptions for many types of goods, it will take a lot of time and effort for people to do by themselves. From that fact, the author wants to contribute a part of his efforts to help reduce the workload of checking customs declarations. In the content of this report, the author will present his solution to the problem of determining product prices through the information described in the customs declaration.

Due to time constraints, in this report, we will focus on the pricing of product descriptions as mobile phones. With automatic price determination accurate enough for the declared description records, the system will help the inspector quickly determine the true value of the item without spending too much time looking up assistance. From there, we can determine the tax rates for each type of goods, supporting fraud detection.

The car industry has become more and more competitive every year and has grown on a global scale. Therefore, it is necessary to set an accurate price for both customers and manufacturers in this competitive car market. Customers and manufacturers are confused about the price of the car to buy or sell. Consequently, customers and manufacturers try to seek some advice from auto-dealers, car magazines or the website on the internet. However, this information takes a long time and might confuse the customers in the market.

N. T. N. Anh et al., *Artificial Intelligence for Automated Pricing Based on Product Descriptions*, SpringerBriefs in Computational Intelligence, https://doi.org/10.1007/978-981-16-4702-4_1

Transportation is a crucial area, therefore, it is unsurprising that the car market plays an important role in developing countries [1]. In Vietnam, where public transport in cities is not fully developed, the main means of transportation are cars and motorbikes. Between cars and motorbikes, cars are gaining popularity across the regions. As a result, the purchase and the sale of cars play a crucial aspect in the economy with a massive scale and various numbers of transactions. With the development of e-commerce, the sale and purchase of cars become much easier between the buyer and the seller. After discovering cars information, we can divide that into two types: structured and unstructured data. Structured data can be divided into two types: numerical and qualitative. However, in order to know the detailed status of the car, it is necessary to look at the unstructured data section in the description, the opinion of people that viewing the information, or the assessments of the seller.

The automotive industry is one of the leading economic sectors in revenue in the world. In particular, in many developing countries, the automobile market has full of potential and is booming strongly as a result of the rapid change in people's needs due to economic development. Consequently, analyzing and predicting car prices in the market will play a crucial role in many practical implications such as analyzing, forecasting and making decisions to sell and buy cars.

Predicting vehicle prices is considered to be a challenging problem because there are many different factors that have an effect on the price of cars. In addition to the characteristics of vehicles such as brands, manufacturers, models, engines, fuel, etc., there are also many external factors that affect the price of cars such as taxes or distance traveled (for used car sales). Some previous studies on predicting car prices are presented in Table 1.1.

From previous studies, it can be seen that the authors chose various factors as input variables to forecast car prices. These features are diverse and consist of many qualitative variables. Therefore, quantifying the qualitative data is a critical step in pre-processing data before putting it into the model to predict vehicle prices. This is also one of the paper's major contributions see Fig. 1.1.

In the real problem, pricing data used websites of selling used cars. The car description data are crawled and priced as in Fig. 1.1.

The objective of this book is to build a model for predicting car prices based on quantitative data analysis and build a knowledge-based system with qualitative data. Furthermore, a new vehicle pricing model will be proposed. First, data is collected from e-commerce sites with many non-structured, digital, non-numeric data fields. After that, the authors store the data to make books. Following that, they develop a method for quantifying qualitative data based on machine learning knowledge. Subsequently, they use the ensemble method to price cars. Finally, the model is applied to the collection of data gathered from the 4 largest automobile trading websites in Vietnam.

Predicting car prices is a regression analysis problem in which a car's price is a dependent variable and the characteristics of the vehicle (brand, car model, registration year, gearbox type, fuel type, ...) are independent variables. We denote the input by $X = \{X^1, X^2, \ldots, X^N\}$ and denote the output by Y. The regression model represents the dependency relationship between Y and X

Table 1.1 Car price prediction researches

Objective	Methodology	Case studies	Year	Pubs
Forecast resale prices of used cars	Using various regression methods	German	2017	[2]
Introducing a new technique	Artificial neural networks, adaptive neuro-fuzzy inference	Taiwan	2009	[3]
Predicting prices of used cars	Random forest	Kaggle	2018	[4]
Predicting prices of used cars in Mauritius	Multiple linear regression, k-nearest neighbors, decision trees, naïve Bayes	Data collected from daily newspapers	2014	[5]
Achieving higher precision	Artificial neural network, support vector machine, random forest	Bosnia and Herzegovina	2019	[6]
Developing prediction system	Multiple linear regression	Pakistan	2017	[7]

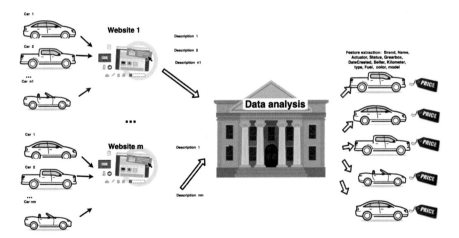

Fig. 1.1 The car description data from many websites are crawled and priced

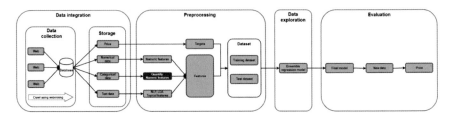

Fig. 1.2   Research method for car pricing problem

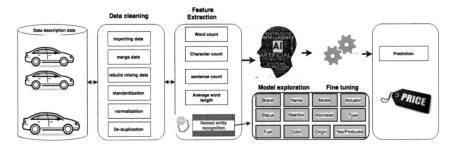

Fig. 1.3   The data set of car description from websites

$$Y = f(X; \theta) \tag{1.1}$$

where f is a function modelling the dependency of the output on the input and parameters θ.

Summary of the steps in the research method for products price predicting problem see Fig. 1.2 is as follows (Fig. 1.3):

- Data integration: integrate data from multiple sources (dataset from crawling from the web, ...)
- Data pre-processing: clean data, fill in the missing data (remove if missing too much information), remove redundant, abnormal, inconsistent data, standardize data, transform data, quantify qualitative data, etc.
- Data exploration: Explore the relationship between the features and the dependent variables to propose models to predict prices.
- Evaluation: Help to find optimal hyper-parameters for the model and choose the most suitable one.

The overview of general products are shown in Fig. 1.2. Specifically, the process to price car has five steps in Fig. 1.2 as follows:

- Data integration: integrate data from multiple sources (data set from crawling from the web, ...)
- Data pre-processing: importing data, merge data, rebuild missing, standardization, normalization, de-duplication.

- Feature extraction: Word count, character count, sentence count, average word length, named entity recognition, quantify qualitative data.
- Data exploration: Explore the relationship between the features and propose models to predict prices.
- Evaluation: fine tuning hyper parameters for the model and predict car price.

1.2 Data Description

1.2.1 Car Description from Customs Declarations

Descriptive data can be obtained from a variety of sources, such as on customs declarations or on e-commerce sites.

> A customs declaration is a document on which the owner of the goods must declare such goods information to the controlling party when importing and exporting goods.

Customs declarations have many forms, with each form consisting of many items, such as cigarettes, telephones, electrical components,... The short description of the items include the information about the goods. To determine the tax value of an item, we need to determine the goods price from its description.

The description of the property is information such as name, manufacturer, new or old status,... The description also includes the features that its manufacturer has disclosed since release such as properties. function, capacity, color,... Thanks to these descriptions one can search for the price of an asset in the market. Examples of some descriptions in practice: " Nokia 625H mobile phone RM-943 CV VN YELLOW—A00013419 (Including body, battery, charger, book, headset, connection cable), 100% new" , " Masstel C105 White mobile phone includes battery body, charger, headset, USB cable and SHD. Features 2 sim, 2 waves, VGA camera, 1.8 in. TFT screen, Bluetooth, FM, 100% new" ,. .. Asset descriptions are unstructured, varied and complex data. So to master them manually will not be easy. Meanwhile, the computer technology and automation industry is growing strongly. So it is necessary and meaningful to automatically handle these complex descriptions automatically (Table 1.2).

1.2.2 Car Description from Websites

The prices of a type of product appear in some specific websites. The prices of used cars in Vietnam are show in many websites sus as vnexpress.net, choxe.net, bonbanh.com, sanoto.com or oto.com.vn. The information is crawled via scrapy that is web scraping, crawling and spiders see Fig. 1.4.

Table 1.2 The descriptions automatically

Classes	Subclass	Length	Action Mechanism
Translation	mRNA[a]	22 (19–25)	Translation repression, mRNA cleavage
Translation	mRNA cleavage	21	mRNA cleavage
Translation	mRNA	21–22	mRNA cleavage
Translation	mRNA	24–26	Histone and DNA modification

[a]Table foot note (with superscript)

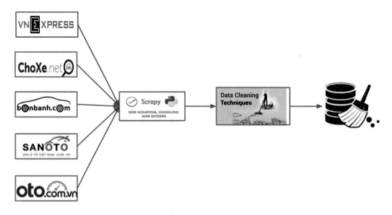

Fig. 1.4 The websites are crawled car description

dan_dong,date,gia,hop_so,km_da_di,loai_xe,may_sac,nam_san_xuat, name,nguoi_ban,nhan_hieu,nhien_lieu,text,tinh_trang,url,xu at_xu
4WD: 4 bánh,29/05/2019,4438000000,Số tự động,0.0,Xe Bán tải,Trắng,2019,F-150 Platium 2019,Đại lý,Ford,Dầu,"Xe trang bị động cơ Ecoboost V6 3.5L, cho công suất 375 mã lực và mô-men xoắn 637 Nm. Hộp số tự động 10 cấp và hệ dẫn động 4 bánh. Ngoài ra còn có một số công nghệ hỗ trợ ra-moóc phía sau. Nội thất của Ford F-150 chẳng khác nào một chiếc SUV hạng sang. Xung quanh vị trí người lái dày đặc nút bấm. Xe có màn hình giải trí cảm ứng kích thước 8 inch, tích hợp hệ thống SYNC 3 và dàn âm thanh Bang & Olufsen 10 loa. Xe trang bị nhiều công nghệ an toàn hiện đại như camera 360, cảnh báo an toàn phía trước, phát hiện người đi bộ, tự động phanh khẩn cấp, cảnh báo chệch làn, hỗ trợ giữ làn, hỗ trợ đổ xe song song, ga tự động thích ứng... Hàng ghế sau rộng rãi, có bệ tỳ tay, hệ thống điều hòa riêng biệt đi kèm một cổng sạc 110V và một cổng sạc 12V. Phần ghế ngồi có thể gập lên để chứa các đồ vật có chiều cao lớn. Xe giao ngay LIÊN HỆ: MINH HÒA GIAO XE TOÀN QUỐC BẢO HÀNH 03 NĂM",Mới,https://choxe.net/xe-moi/ban-xe-moi-ford-f-150-platium-2019-tai-ha-noi-316717.html,Nhập khẩu
RWD: Cầu sau,30/04/2019,505000000,Số sàn,28000.0,Xe Bán tải,Đỏ,2017,Colorado 2.5L 4x2 MT 2017,Đại lý,Chevrolet,Dầu,"Salon ô tô Ánh Lý Chevrolet Colorado 2.5LT dòng bán tải Mỹ nhập khẩu nguyên chiếc sàn xuất 2017 biển tỉnh hồ sơ rút nhanh gọn. ĐẶC ĐIỂM: Động cơ Duramax tăng áp 2.5L Công suất cực đại 161 Hp Nội thất, ni cao cấp Hộp số sàn 6 cấp Một cầu chủ động Màn hình cảm ứng Mylink 7 inch kết nối Apple Carplay, 4 loa, AUX, USB, kết nối Bluetooth Động cơ tăng áp Duramax 2.5L trang bị cho phiên bản LT cho công suất cực đại 161 mã lực tại 3.600 vòng/phút, và mô men xoắn cực đại là 380 Nm tại 2.000 vòng/phút. Hỗ trợ trả góp lãi xuất ưu đãi thủ tục nhanh gọn.",Cũ,https://choxe.net/xe-cu/ban-xe-cu-chevrolet-colorado-25l-4x2-mt-2017-tai-phu-tho-308322.html,Nhập khẩu
RWD: Cầu sau,16/05/2019,605000000,Số tự động,22000.0,Xe Bán tải,Xanh,2017,Ranger 2.2AT 2017,Đại lý,Ford,Dầu,"Salon ô tô Ánh Lý bán xe Ranger 2.2AT màu xanh xám sàn xuất 2017 đăng ký 2018 hồ sơ rút ngay trong ngày. Xe còn như mới, phiên bản đủ đồ: Số tự động,1 cầu, mâm đúc, ghế da, vô lăng tích hợp phím điều chỉnh âm thanh, đầu đĩa DVD kết hợp cảm biến lùi. Cam kết chất lượng: xe không đâm đụng không ngập nước máy nguyên bản của nhà xuất. Hỗ trợ trả góp lên đến 70% lãi xuất ưu đãi thủ tục nhanh gọn giao xe ngay. Vui lòng liên hệ chúng tôi để biết thêm thông tin chi tiết!",Cũ,https://choxe.net/xe-cu/ban-xe-cu-ford-ranger-22at-2017-tai-phu-tho-314110.html,Nhập khẩu
FWD: Cầu trước,05/09/2018,1199000000,Số tự động,0.0,SUV,Nâu,2018,3008 All New 2018,Đại lý,Peugeot,Xăng,"Mẫu xe Peugeot

Fig. 1.5 Example of the raw data collected from websites

dan_dong	gia	hop_so	km_da_di	loai_xe	mau_sac	nhan_hieu	nhien_lieu	xuat_xu	model	phien_ban	old	xe_model	xe_model_phien_ban
RWD	505000000	Số tay	28000	Truck	Đỏ	chevrolet	Dầu	Nhập khẩu	colorado	-		2.326027367	chevrolet colorai -
RWD	605000000	Số tự động	22000	Truck	Xanh	ford	Dầu	Nhập khẩu	ranger	-		2.369863014	ford ranger
FWD	435000000	Số tự động	39500	Sedan	Trắng	hyundai	Xăng	Trong nước	avante	1.6 at		4.994520548	hyundai avante hyundai avante 1.6 at
FWD	595000000	Số tự động	21000	Sedan	Trắng	honda	Xăng	Trong nước	city	top		1.408219178	honda city honda city top
FWD	485000000	Số tự động	38000	Sedan	Bạc	chevrolet	Xăng	Trong nước	cruze	-		4.147945205	chevrolet cruze -
FWD	298000000	Số tay	100000	Sedan	Đen	honda	Xăng	Trong nước	civic	1.8mt		10.4109589	honda civic honda civic 1.8mt
FWD	475000000	Số tự động	80000	SUV	Xám	honda	Xăng	Trong nước	cr-v	at		10.4109589	honda cr-v honda cr-v at
FWD	635000000	Số tự động	20000	Sedan	Trắng	kia	Xăng	Trong nước	cerato	2.0at		1.408219178	kia cerato kia cerato 2.0at
FWD	170000000	Số tự động	120000	Hatchback	Đỏ	kia	Xăng	Nhập khẩu	morning	six		12.41369863	kia morning kia morning six
RWD	1100000000	Số tự động	32000	Sedan	Đỏ	bmw	Xăng	Nhập khẩu		1 -		4.408219178	bmw 1 bmw 1 -
RWD	485000000	Số tay	67000	Truck	Đỏ	ford	Dầu	Nhập khẩu	ranger	-		6.408219178	ford ranger -
FWD	205000000	Số tay	100000	Sedan	Đen	chevrolet	Xăng	Trong nước	aveo	mt		6.408219178	chevrolet aveo chevrolet aveo mt
FWD	320000000	Số tay	110000	Sedan	Cát cháy	chevrolet	Xăng	Trong nước	cruze	1.6mt		8.410958904	chevrolet cruze chevrolet cruze 1.6mt
FWD	630000000	Số tự động	40000	Sedan	Trắng	mazda	Xăng	Trong nước		2 -		4.408219178	mazda 2 mazda 2 -
RWD	775000000	Số tự động	50000	SUV	Trắng	toyota	Xăng	Trong nước	fortuner	at		5.408219178	toyota fortuner toyota fortuner at
FWD	480000000	Số tự động	80000	SUV	Xám	honda	Xăng	Trong nước	cr-v	at		10.4109589	honda cr-v honda cr-v at
FWD	585000000	Số tự động	40000	Sedan	Đen	mazda	Xăng	Trong nước		2 3 at 015		3.408219178	mazda 2 3 at 015
4WD	435000000	Số tự động	120000	SUV	Xanh	hyundai	Dầu	Trong nước	santa fe	at		12.41369863	hyundai santa fe hyundai santa fe at
FWD	179000000	Số tự động	80000	Sedan	Xám	morning	Xăng	Trong nước	morning	sx		8.410958904	kia morning kia morning sx
4WD	189000000	Số tự động	170000	SUV	Xám	ford	Xăng	Trong nước	escape	-		17.41643836	ford escape
FWD	295000000	Số tay	80000	Sedan	Xám	daewoo	Xăng	Trong nước	lacetti	1.6mt		9.410958904	daewoo lacetti daewoo lacetti 1.6mt
FWD	255000000	Số tay	64625	Hatchback	Trắng	kia	Xăng	Trong nước	morning	1.25mt		3.408219178	kia morning kia morning 1.25mt
FWD	630000000	Số tự động	40000	Sedan	Trắng	mazda	Xăng	Trong nước		2 6 .0 at 015		4.408219178	mazda 2 mazda 2 6 .0 at 015
RWD	275000000	Số tay	50000	Pick-up Truck	Trắng	kia	Dầu	Trong nước	k	-		3.408219178	kia k kia k -
FWD	720000000	Số tự động	70000	SUV	Đen	toyota	Xăng	Nhập khẩu	venza	-		10.4109589	toyota venza -

Fig. 1.6 The examples of data collected from websites are clean

The news of used car in car market websites are provided by sellers and buyer. The raw data collected from websites are save in one data base that is updated by day Fig. 1.5. The model of extraction data collected from websites is clean see Fig. 1.6.

The data is the new oil of the digital economy. Pricing product is interesting problem to exploit this new oil.

References

1. Kumar B, Sarkar P (2016) Prediction of future car forms based on historical trends. Perspect Sci Recent Trends Eng Mater Sci 8:764–766. ISSN 2213-0209, https://doi.org/10.1016/j.pisc. 2016.06.082
2. Lessmann S, Voß S (2017) Car resale price forecasting: The impact of regression method, private information, and heterogeneity on forecast accuracy. Int J Forecasting 33(4):64–877. ISSN 0169-2070, https://doi.org/10.1016/j.ijforecast.2017.04.003
3. Wu J-D, Hsu C-C, Chen H-C (2009) An expert system of price forecasting for used cars using adaptive neuro-fuzzy inference. Expert Syst Appl 36(4):7809–7817. ISSN 0957-4174, https://doi.org/10.1016/j.eswa.2008.11.019
4. Pal N, Arora P, Sundararaman D, Kohli PS, Palakurthy S (2017) How much is my car worth? A methodology for predicting used cars prices using Random Forest.
5. Pudaruth S (2014) Predicting the price of used cars using machine learning techniques. Int J Inf Comput Technol 4:753–764
6. Gegic Enis (2019) Car price prediction using machine learning techniques. TEM J 8(1):113–118
7. Noor K, Jan S (2017) Vehicle price prediction system using machine learning techniques. Int J Comput Appl 167:27–31 https://doi.org/10.5120/ijca2017914373

Chapter 2
Machine Learning and Ensemble Methods

2.1 Introduction to Machine Learning

Herbert Alexander Simon was archived Turning Award in 1995 and Nobel prize in economics in 1978 defined learning as any process by which a system improves performance from experience [2]. Then, machine learning mention with computer programs that ability to learn without being explicitly program that improves their performance from experience.

Machine learning discovers new knowledge from big data collected via activities such as applications from mobile, social data, websites. Using machine learning help develop automatically adapt and customize the system by changing of data of individual users. Machine learning help the system recognizing that replace human such as faces, objects, handwriting characters, voice.

Machine learning is separated in three types: Supervised, unsupervised and reinforcement. Firstly, supervised learning includes classification and regression problems that output of classification is a discrete variable and output of regression is continuous variable. Secondly, Unsupervised learning is no desired output that learn something from data and latent relationships of data such as clustering that learns structure in the data. Finally, reinforcement learning defined an agent interacts with an environment and receives feedback reward signal.

The processes solving a machine learning problem are common five steps:

1. Data gathering: Collect data from various sources. Data is collected depend on human work such as manual labeling for supervised learning or experts suggest domain knowledge.
2. Data pre-processing: Clean data to have homogeneity. The data need clean such as missing values, outliers, bad encoding, wrongly-label, biased data.
3. Feature engineering: Making data more useful input of machine learning is a set of features. Feature engineering extracts more information from existing data that require knowledge of the data such as variable transformation (e.g. date into weekdays, normalizing), derived features (e.g. n-gram text, ratio).

N. T. N. Anh et al., *Artificial Intelligence for Automated Pricing Based on Product Descriptions*, SpringerBriefs in Computational Intelligence, https://doi.org/10.1007/978-981-16-4702-4_2

4. Modeling: Select the single or ensemble machine learning models. Supervised learning model such as linear classifier, Naive Bayes, Support vector machines, Decision tree, Random forests, neural networks and k-nearest neighbors; Unsupervised learning such as PCA, k-mean, SBSCAN; Reinforcement learning such as Q-learning or SARSA.
5. Evaluation and deploy: Evaluate the models and deploy. This step uses of metrics for evaluating performance and comparing methods. This step includes tuning hyper-parameters and then deploy.

2.2 Linear Classification

The goal of linear classification model is that separates the input data into different regions. Concretely, the given input data $X_i = \{x_i^1, \ldots, x_i^m\}$ where $i = 1, \ldots, N$ instances, assign each instance to one of K discrete classes C_k where $k = 1, \ldots, K$. The class labels are discrete. Two types of classification are binary classification if $K = 2$ and multi-class if $K > 2$.

Linear model as in linear regression that $y(X_i, w)$ is linear function of the parameter w:

$$y(X_i, w) = w^T \phi(X_i)$$

where $y(X_i, w) \in \mathbb{R}$.

Classification model has a discrete label so a mapping is applied to the linear model $f(.) : \mathbb{R} \to \mathbb{Z}$

$$y(X_i, w) = f(w^T \phi(X_i))$$

To functions $f(.)$, $f^{-1}(.)$ are called activation function and link function respectively. $f(.)$ is the essential problem of classification problem that has three methods:

1. Find a discriminant function $f(.)$ that maps each input data into a class label
2. Discriminate models that solve the inference problem to determining the posterior probability $Pr(C_k|X_i)$. Then use decision theory to determine new x_j to one of the class C_l.
3. Generative models that solve the inference problem to determining the given class conditional probability $Pr(X_i|C_k)$. Then infer the probability $Pr(C_k)$. Next, applying Bayes's theorem to calculate the $Pr(C_k|x)$. Then, model the join distribution $p(X_i, C_K)$ is use to assign each new x_j to one of the class C_l by decision theory.

Example 2.1 A discriminant function is maps from an input vector X_j to on of $K = 2$ classes. The linear function of the input X_j

$$y(X_j) = w^T X_j + w_0$$

where w is weight vector and w_0 is bias.

Example X_j is assigned to class C_1 if $y(X_j) \geq 0$ and C_2 otherwise.

The decision boundary $y(X_j) = 0$ is decision surface and w is orthogonal to any vector in the decision surface. The normal distance form the origin to the decision surface is

$$\frac{w^T X_i}{||w||} = -\frac{w_0}{||w||}$$

Let $X_{j,project}$ be the projection point of X_j on the decision surface, so $w^T x_{project} + w_0 = 0$, and d be the distance from X_j to the decision surface. We have:

$$y(X_j) = w^T \left(X_{j,project} + d\frac{w}{||w||} \right) + w_0 = d\frac{w^T w}{||w||} + w^T X_{j,project} + w_0 = d||w||.$$

Thus, $d = \dfrac{y(X_j)}{||w||}$

The multi-class has $K > 2$ that binary class discriminant functions using $K - 1$ one-versus-the-rest or $K(K - 1)/2$ one-versus-one classifiers. K functions $y_k(X_j) = w_k^T X_j + w_{k0}$ the instance X_j is assigned to class C_k if $y_k(X_j) \geq y_l(X_j)$; $\forall l \neq k$.

The metric valuation of classification is the sum square error function. The available methods are used in classification problems such as least square error or Fisher's linear discriminant.

2.3 Support Vector Machine

Support Vector Machine (SVM) is a strong tool for classifier model that is related to statistical learning theory. The main objective of SVM is a discriminate classifier that intakes training data and an optimal hyper-plane. SVM is avoid over-fitting with very high dimensional features of data. Support vectors are data points that are close to the decision surface which difficult to classify. This method is over the barrier of optimum location on the surface.

SVM is common because of its success in handwritten digit recognition and the same error rates with the neural network. In addition, SVM is kernel method that is the key in machine learning.

Given training data (X_j, y_j), $j = 1, \ldots, n$ and $y_j \in \{-1, 1\}$. The learning classifier $f(X_j)$ is defined as follows:

$$y_j = \begin{cases} 1 & \text{if } f(X_j) \geq 0; \\ -1 & \text{if } f(X_j) < 0; \end{cases} \tag{2.1}$$

The $y_j.f(X_j) > 0$ if the classification is correct. The linear classifiers has the form

$$y(X_j) = w^T X_j + w_0$$

where w is the weight vector and w_0 is the bias. The Perceptron algorithm:

$$f(X_j) = w^T X_j + w_0 = \tilde{w}^T \tilde{X}_j$$

where $\tilde{w} = (w, w_0)$, $\tilde{X}_j = (x_j, 1)$.

- Initial $\tilde{w} = 0$
- Cycle though the data instance (X_j, y_j)
  if X_j is misclassified then $\tilde{w} \leftarrow \tilde{w} + \alpha sign(f(X_j)).X_j$
- Until all the data classified are correct.

SVM is prefer a larger margin for generalization so hyperplan H is defined
$w^T X_j + w_0 \geq 1$ when $y_j = 1$
$w^T X_j + w_0 \leq -1$ when $y_j = -1$
H_1 plane is $w^T X_j + w_0 = 1$
H_2 plane is $w^T X_j + w_0 = -1$
The instances on the planes H_1, H_2 are the support vectors.
$d^+ = $ is the shortest distance from the hyper-plane to the closet positive point,
$d^- = $ is the shortest distance from the hyper-plane to the closet negative point.
The margin of separating hyper-plane H is $d^+ + d^-$. The purpose of SVM classifier is maximum the margin.
The distance of H and H_1 is

$$\frac{w^T X_j + w_0}{||w||} = \frac{1}{||w||}$$

Therefore, the distance of H_1, H_2 is $d^+ + d^- = \dfrac{1}{||w||}$. Thus, maximize the margin
is minimize $||w||$. There are not any instance data between H_1, H_2. We see that
$w^T X_j + w_0 \geq 1$ when $y_j = 1$ and $w^T X_j + w_0 \leq -1$ when $y_j = -1$ so $y_j.f(X_j) \geq 1$.

SVM is applied Lagrangian method to maximize (Figs. 2.1 and 2.2)

$$L(w, w_0, \alpha) = \frac{1}{2}||w|| - \sum_{j=1}^{n} \alpha_j[y_j.(w^T X_j + w_0) - 1] \qquad (2.2)$$

The partial derivative of L with variable w, w_0, are equal 0 we have:
Maximize $\sum_{j=1}^{n} \alpha_j - \frac{1}{2} \sum_{i,j} y_i y_j \alpha_i.\alpha_j < X_i, X_j >$.
Subject to $\sum_{j=1}^{n} y_j \alpha_j = 0$ and $\alpha_j \geq 0$.
The SVM is convert to quadratic programming that can solved in $O(n log n)$ time.
When the decision function is nonlinear, the kernel method is applied.

Fig. 2.1 Support vector machine with the hyperlane

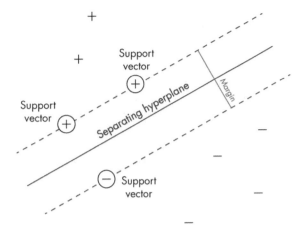

Fig. 2.2 Ensemble methods

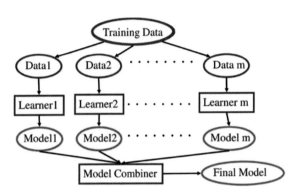

Maximize $\sum_{j=1}^{n} \alpha_j - \frac{1}{2} \sum_{i,j} y_i y_j \alpha_i . \alpha_j < F(X_i), F(X_j) >$.

Subject to $\sum_{j=1}^{n} y_j \alpha_j = 0$ and $\alpha_j \geq 0$.

The kernel $K(X_i, X_j) = < F(X_i), F(X_j) >$

SVM is a useful and common in machine learning that has been appeared in many applications.

2.4 Decision Tree

A Decision tree is a structure that divides the data of instances into smaller sets of instances by a sequence of simple decision rulers. The elements of the resulting sets become similar to the other elements of the same set. Concretely, a decision tree model includes a set of rulers for separating large heterogeneous instances into smaller that more homogeneous classes respecting a specific target. Therefore, the decision tree algorithm is a recursive partitioning algorithm.

A decision tree has nodes, leaves and branches. There are three types of nodes that are: (i) root node in the to node with no incoming edges but outgoing edges. (ii) Internal node has exactly one incoming edge and two or more outgoing edges; (iii) Leaf node is a terminal node that has one incoming edge but no outgoing edges. Leaf node is assigned a class label and branches is unique edges with a set of attribute that separate the observations into smaller.

Some notations are used for the decision tree models to describe instances, attributes, the class labels and the tree structure. An instance X_j is a vector of m attribute values with an target class y_j. $\mathbf{A} = \{A_1, A_2, \ldots, A_m\}$ is the set of attributes of data. Thus, the instance $X_j = \{x_j^1, \ldots, x_j^m\}$ where $x_j^1 \in A_1, \ldots, x_j^d \in A_d$. $\mathbf{Y} = \{y_1, \ldots, y_N\}$ be the set of class labels. The number instances in $\mathbf{X}^l$ that belong to class y_j is X_j^l. The probability of selection randomly the instance X^l is of class y_j is $p_{lj} = \dfrac{|X_j^l|}{|X^l|}$.

The metric to measure of misclassified instances. If y_j is the class value that occurs in partition X^l is

$$Error(X^l) = E(X^l) = \frac{|\{y \neq y_j : (X_q, y) \in X^l\}|}{|X^l|} = 1 - p_{lj}.$$

Thus, the error rate of the total X is calculate by the below formula:

$$Error(X) = E(X) - \sum_{l=1}^{K} Error(X^l) . \frac{|X^l|}{|X|}$$

The metric evaluation of decision tree model is entropy and information gain formulated as

$$H_X(Y) = - \sum_{y \in Y} p_y log p_y$$

where p_y is the probability of random selection label y. Thus, the information entropy of all tree that weights sum of the entropies for each partition $H_{X^l}(Y)$ formulated as

$$Gain(X, A) = H_X(Y) - \sum_{l=1}^{K} \frac{|X^l|}{|X|} . H_{X^l}(Y^l).$$

Gini index is used for a splitting criterion.

$$Gini(X) = 1 - \sum_{i=1}^{K} p_l^2.$$

ID3, C4.5, C5.0 are considered the common decision tree algorithms.

The advantages of a decision tree are easy to interpret, handling both numerical and categorical variables, training time is fast being compared with other machine learning model and learning incrementally. However, the disadvantages of this model are not good for multivariate partitions, focusing on the training data referring easily over-fit and sensitivity with data diversity.

To reduce the advantage of a decision tree model, the random-forest is considered. An ensemble classifier using many decision tree models that used for classification or regression are called random forests. This model has out-perform and variable importance information than a decision tree. The random forest model can handle the over-fitting of individual trees' predictions [1].

The idea of a random forest algorithm is chosen the number m_r of input variables (m_r is much less than m). The m_r variables determined decision at a note of the tree. Then, a training set for this tree by choosing n times with replacement form n instances of the training data set. The test data is used to evaluate the error of this tree based on the prediction results. For each node of the tree, randomly chose m_r variable and calculate the best split based on there m_r variable in the training data set. We see that each tree is fully grown and not pruned.

2.5 Ensemble Methods

Ensemble methods are learned multiple alternative models using different training data or different learning models that combine decision of multiple modes using weight vectors. Ensemble methods enhance the performance of a single machine learning model.

The methods of ensemble learning are Voting and averaging, stacking, Bagging and Boosting.

- The voting is used in classification and the averaging is used in the regression.
- Stacking is using different learning models and then combined them together by another machine learning algorithm called a combiner. The learning models generate a new data set. The combiner is used this new data set for input.
- Bagging: from the original data set with n instances. The sub-sample is generated by the bootstrap sampling method. Then, multiple machine learning models are applied for sub-sample. Next, voting or averaging are regarded for the final results.
- Boosting: sequentially models with the performance of the prior model will use to build the next one. Performance of boosting based on the Weights are allocated to each model. The boosting algorithms have out-performance than a single model with low variance and low bias. Boosting model is a power that runs fast and stable over time.

Example 2.2 Voting is considered the most simple ensemble approach because it is easy to understand and implement. Firstly, the whole training data is used to m classifier models. Secondly, the final result is decided by all results of a single model having the same voting power that sees Fig. 2.3. This example, SVM, KNN, LR,

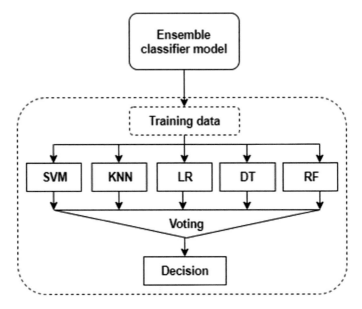

Fig. 2.3 Voting ensemble methods used SVM, KNN, LR, DT, RF as single base classifiers

DT, RF are single base classifiers to be applied to pricing product problems. Voting ensemble classifications are constructed from single results. Ensemble model face with over-fitting the training data than single base model. Thus, choosing an ensemble model overpass over-fitting is an interesting problem. The weighted votes of weak classifiers are summaries for the final classifier.

XGBoost, AdaBoots CatBoost and LightGBM are common algorithm for boosting algorithm.

Example 2.3 In this chapter, Adaboots that notation for the Adaptive booosting approach is chosen to consider for represent the example of boosting approach. This model is constructing classifier by combination weak classifiers (Fig. 2.4)

$$f(x) = \sum_{i=1}^{n} \alpha_i h_i(x)$$

The code python for classification iris data that is common in classification problem using a linear model as follows:

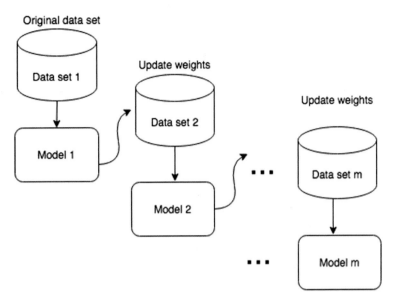

Fig. 2.4 Adaboost methods is sequentially models with the performance of the prior model will use to update weighted the next one

Linear model

```
import pandas as pd
from sklearn import datasets
import sklearn as sk
from sklearn.neural_network import MLPClassifier
from sklearn.model_selection import train_test_split
from sklearn import svm
from sklearn.linear_model import LogisticRegression
from sklearn.ensemble import RandomForestClassifier


# import some data to play with
iris = datasets.load_iris()
X = iris.data
y = iris.target

X_train, X_test, y_train, y_test =

train_test_split(X, y, test_size=0.3, random_state=0)
```

```
LinReg = LogisticRegression(random_state=0, solver='lbfgs',

multi_class='multinomial').fit(X_train, y_train)
LinReg.predict(X_test)
print(' Linear regression:',
round(LinReg.score(X_test,y_test), 4))

SuppVM = svm.SVC(decision_function_shape="ovo").

fit(X_tr, y_tr)
SuppVM.predict(X_test)
print(' SVM:',round(SuppVM.score(X_test,y_test), 4))

RandF = RandomForestClassifier(n_estimators=1000,

max_depth=10,

random_state=0).fit(X_train, y_train)
RandF.predict(X_test)
print(' Random Forest:',round(RF.score(X_test,y_test), 4))

neural_network = MLPClassifier(solver='lbfgs', alpha=1e-5,

hidden_layer_sizes=(150, 10), random_state=1).
fit(X_train, y_train)
neural_network.predict(X_test)
round(neural_network.score(X_test, y_test), 4)
print(' MLPClassifier:',round(NN.score(X_test,y_test), 4))
```

References

1. Pal N, Arora P, Sundararaman D, Kohli PS, Palakurthy S (2017) How much is my car worth? A methodology for predicting used cars prices using Random Forest. https://ui.adsabs.harvard.edu/abs/2017arXiv171106970P
2. Richard NL (2003) Cognitive comparative advantage and the organization of work: lessons from Herbert Simon's vision of the future. J Econ Psychol 24(2):167–187. https://doi.org/10.1016/S0167-4870(02)00201-5

Chapter 3
Quantifying the Qualitative Features

3.1 Qualitative Features

Machine learning and deep learning archive significant advances in the past few decades [1]. The data scientists and researchers in machine learning have faced six challenges: Data collection, less amount training data, non-representative training data, poor quality of data, irrelevant features, over-fitting/ under-fitting the training data. In machine learning, the decision tree algorithm can be trained directly from categorical data. However, the other machine learning algorithms can not handle with categorical data (qualitative features) directly. Categorical variables require transform into be numeric variables.

A categorical variable is a variable represented qualitative property of an object. Categorical variables derive from observations made of qualitative data. In data science, machine learning needs to handle categorical variables.

One of the problem of the non-representative data is quantitative categorical variables. Qualitative categorical variables is essential pre-processing data because of providing insights information data into the next step of the training model.

Quantifying the datatype is vital analyzing quality data. The machine learning models or deep learning models will not be well learned if category variables are not well represented [2].

Categorical variables represent types of data of a product such as a form, shape, line, color, space and texture, branch. The car product the categorical variables are: gearbox, seller, type, fuel, color and origin. Several methods are given to solve categorical variables as follows:

- Integer encoding
- One-hot encoding
- Entity embedding.

N. T. N. Anh et al., *Artificial Intelligence for Automated Pricing Based on Product Descriptions*, SpringerBriefs in Computational Intelligence, https://doi.org/10.1007/978-981-16-4702-4_3

3.2 Integer Encoding

The simple method to handle with categorical variable is integer encoding. The categorical values are converted to integer values.

Example 3.1 Example a car is evaluated four bands $Toyota, Bentley, BMW, Tesla$ are changed to their corresponding integer:

$Toyota$ is 2,
$Bentley$ is 0,
$Tesla$ is 1.

However, $0 < 1 < 2$, the numbers have order thus it is not true in representing.

Using python to encode the categorical data into integer encoding as follows: A data frame of brand and price of cars are created:

Create data of cars

```
import pandas as pd
import numpy as np

cars = {'Brand': ['Toyota','Bentley','Tesla'],
        'Price': [2200,105000,37000]
        }

df = pd.DataFrame(cars, columns = ['Brand', 'Price'])

print (df)

     Brand    Price
0   Toyota     2200
1  Bentley   105000
2    Tesla    37000
```

The types of car data considered:

types of car data

```
df.dtypes
```

```
Brand       object
Price        int64
dtype: object
```

Changing the object type to category type by this command:

Change the object tyoe

```
df['Brand'] = df['Brand'].astype('category')
df.dtypes
```

```
Brand     category
Price        int64
dtype: object
```

The command cat.codes change categorical data into integer encoding:

change categorical data into integer

```
df['Brand_cat'] = df['Brand'].cat.codes
print(df)
```

```
       Brand    Price  Brand_cat
0     Toyota     2200          2
1    Bentley   105000          0
2      Tesla    37000          1
```

Example 3.2 The gender feature of a person is categorical variables with two values "male" , "female" that is integer encoding:

male is 0

female is 1

Using python to encode the categorical data into integer encoding as follows:

Categorical data into integer encoding example

```
import pandas as pd
import numpy as np

people = {'Name': ['Dung','Hang', 'Hung', 'Linh'],
          'Gender': ['male','female','male','female', ]
          }

df = pd.DataFrame(people, columns = ['Name', 'Gender'])
print(df)

   Name  Gender
0  Dung    male
1  Hang  female
2  Hung    male
3  Linh  female

df['Gender'] = df['Gender'].astype('category')
df['Gender'] = df['Gender'].cat.codes
print(df)

   Name  Gender
0  Dung       1
1  Hang       0
2  Hung       1
3  Linh       0
```

However, $0 < 1$ is not good in representing.

The integer encoding has the advantage in that it is easily reversible the original value of categorical variables. The integer values have ordered, the machine learning train the value and relationship of the volume of integers. Thus Integer encoding of categorical variable is not a good presentation.

3.3 One-Hot Encoding

One method to convert category data into a new column and assigns a 1 or 0 is one hot encoding. The advantage of this method is not weighting a value improperly but it has the downside of adding some columns into the data set. We reconsider the Example 3.1 and Example 3.2, the categorical variable is used *get_dummies* to convert to columns with 0 or 1 corresponding to the true value as follows:

Example 3.3

Data car example

```
import pandas as pd
import numpy as np

cars = {'Brand': ['Toyota','Bentley','Tesla'],
        'Price': [2200,105000,37000]
        }

df1 = pd.DataFrame(cars, columns = ['Brand', 'Price'])

print (df1)
 Brand    Price
0   Toyota    2200
1   Bentley  105000
2    Tesla   37000

df1=pd.get_dummies(df1, columns=['Brand'])
print(df1)
Price  Brand_Bentley  Brand_Tesla  Brand_Toyota
0    2200             0            0             1
1  105000             1            0             0
2   37000             0            1             0
```

Example 3.4

Create people data example

```
import pandas as pd
import numpy as np

people = {'Name': ['Dung','Hang', 'Hung', 'Linh'],
        'Gender': ['male','female','male','female', ]
        }

df = pd.DataFrame(people, columns = ['Name', 'Gender'])
print(df)
```

```
df=pd.get_dummies(df, columns=['Gender'])
print(df)
```

The alternative python one hot encoder using sklearn package using OneHotEncoder() function to convert categorical variable to 0 or 1 values. Then, the new one hot encoder columns will join the original data set.

Example 3.5

One-hot Encoder example

```
from sklearn.preprocessing import OneHotEncoder
import pandas as pd
import numpy as np

cars = {'Brand': ['Toyota','Bentley','Tesla','Toyota'],
        'Price': [2200,105000,37000, 2100]
        }

df1 = pd.DataFrame(cars, columns = ['Brand', 'Price'])

oe_style = OneHotEncoder()
oe_results1 = oe_style.fit_transform(df1[['Brand']])
pd.DataFrame(oe_results1.toarray(), columns=oe_style.
categories_).head()
df1=df1.join(pd.DataFrame(oe_results1.toarray(),
columns=oe_style.categories_))
print(df1)

      Brand    Price   (Bentley,)   (Tesla,)   (Toyota,)
0    Toyota     2200          0.0        0.0         1.0
1   Bentley   105000          1.0        0.0         0.0
2     Tesla    37000          0.0        1.0         0.0
3    Toyota     2100          0.0        0.0         1.0
```

The disadvantage of one hot encoding is the high dimensional and sparse if data have very many unique values in a column and the relations between different values of categorical variables are ignored [3].

3.4   Entity Embedding with Neural Networks

Entity embedding learns supervised representations of arbitrary categorical variables. The concept of an entity embedding is a mapping of a categorical variable into an n-dimensional vector. This method is a helpful tool for solving categorical data. Entity embedding reduces the memory usage by decreasing the features comparing one hot encoder. In addition, this method improves the performance of the neural network because of better data representation and groups similar variables together [3] (Table 3.1).

Some studies of quantifying qualitative data are presented in Table 3.2.

3.4.1   Neural Networks

A neural network (NN) is a method of computing that interacts with multiple connected nodes that are processing elements. NN has three types of layers that are input layer, hidden layers and output layer. Each layer has the number of nodes with the task received input, process and obtain an output; the links these nodes from the previous layer to next layer. The number of hidden layers and the number of nodes in each layer are hyperparameters of neural networks. In addition, the activation function in neural network terminology is necessary to define. The common activations are used such as relu, signmoid and tanh [8] (Fig. 3.1).

3.4.2   Entity Embedding

The entity embedding (EE) of categorical variables is a map of categorical variables into Euclidean space. The neural network helps us to find the map (Figs. 3.2 and 3.3).

Table 3.1 Researches on quantify qualitative data

Objective problem	Methodology	Case studies	Year	Pubs
Quantify qualitative data	Entity Embedding	Rossmann Store Sales	2016	[3]
Dummy coding vs effects coding	ANOVA, dummy coding, effects coding	Dutch online panel	2016	[4]
Improving the efficacy of the lean index	Balanced scorecard, Benchmarking, Strategos LAT		2015	[5]
Class noise cleaning	Clear noise label by ensemble filtering and noise scoring	31 sets of classified data taken from the UCI [6]	2017	[7]

Table 3.2 The features of car are produce from the raw data of these websites

No	Types	Name	Description
1	Categorical	Brand	The brand of manufacturing company
2	Number	Price(VND)	Price that owner proposes to sell the car
3	Text	Name	Name of car
4	Categorical	Actuator	FDW, AWD, etc.
5	Categorical	Status	New, old
6	Categorical	Gearbox	automatik, manuell
7	Datetime	DateCreated	The date at web was created
8	Categorical	Seller	Private or dealer
9	Number	Kilometer	A number of kilometers driven
10	Number	YearProduced	Year the car was produced
11	Categorical	type	a small car, bus, limousine, etc.
12	Categorical	fuel	benzin, Diesel,etc.
13	Categorical	color	Color of car
14	Text	description	Description of car
15	Categorical	origin	domestic, Import

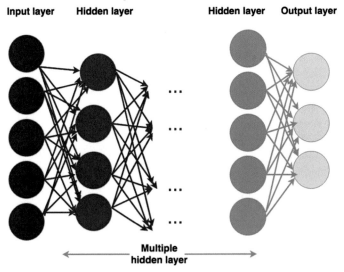

Fig. 3.1 Neural network has three types of layers that are input layer, hidden layers and output layer

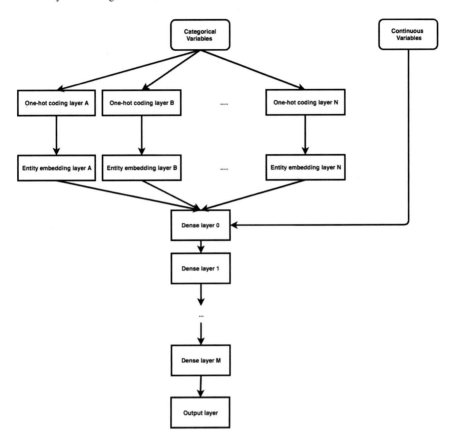

Fig. 3.2 Entity embedding

Fig. 3.3 Embedding Layer

	One-hot encoding								Embedding layer			
Mon	1	0	0	0	0	0	0		0.62	0.71	0.56	0.81
Tue	0	1	0	0	0	0	0		0.28	0.20	0.36	0.45
Web	0	0	1	0	0	0	0		0.52	0.92	0.91	0.54
Thu	0	0	0	1	0	0	0	x	0.06	0.72	0.18	0.30
Fri	0	0	0	0	1	0	0		0.76	0.41	0.41	0.32
Sat	0	0	0	0	0	1	0		0.26	0.69	0.43	0.62
Sun	0	0	0	0	0	0	1		0.59	0.17	0.64	0.56
			7x7								7x4	

From the one-hot encoded matrix, the one-hot encoded matrix is an identity matrix is combined with the entity embedding matrix to make the same entity embedding matrix. The neural net will be used this matrix to updated during the back-propagation process to get the real number matrix.

The python code of the entity embedding method to convert categorical data to real number data.

Example 3.6

Entity embedding code example

```
x_embedded = np.ones((tempt.shape[0], 1))

for idx, cat in enumerate(cat_vars):
    emb_matrix = embs[idx]
    tempt[cat] = tempt[cat].astype('category').cat.as_ordered()
    mapped = np.vstack(tempt[cat].cat.codes.map(lambda i:
    emb_matrix[i+1]))
    print(mapped.shape)
    x_embedded = np.hstack([x_embedded, mapped])

for cont in cont_vars:
    val = tempt[cont].values.reshape((tempt.shape[0],1))
    x_embedded = np.hstack([x_embedded, val])


emb_final = {}
for idx, cat in enumerate(cat_vars):
    emb_final[cat] = {}
    keys = tempt[cat].unique().sort_values()
    for idy, key in enumerate(keys):
        emb_final[cat][key] = embs[idx][idy + 1].tolist()
```

Entity embedding converts categorical to real numbers handle the disadvantage of integer encoding and one hot encoding. This method is used widely in machine learning change qualitative data to quantitative data.

Table 3.3 Applied the proposed model (entity embedding and ensemble learning) with others using Vietnamese data set

No	Model	R^2
1	XGBoost	0.6497
2	Entity embedding and XGBoost	**0.7735**
3	Random Forest	0.7280
4	Entity embedding and RandomForest	**0.7826**
5	LightGBM	0.7060
6	Entity embedding and LightGBM	**0.8270**

3.5 Applications

We used Vietnamese car data set that was collected from 4 commercial websites and newspaper pages choxe.net, sanotovietnam.com, bonbanh.com, and oto.com.vn. These websites are famous and transaction old cars between customers of old car market. The Vietnamese data set is crawled by scrapy framework. Fifteenth features of cars are produced from the raw data of these websites that enough for us to price by machine learning.

The fifteen features having eight categorical features that are Brand, Name, Actuator, Gearbox, Seller, Type, Color and Origin that need to transform to the real number features by applied EE.

Before the Vietnamese data set are converted to quantification, then RandomForest, XGBoost, and LightGBM are used to predict car prices (Table 3.3).

The proposed model combining entity embedding and ensemble learning such as XGBoost, RandomForest and LightGBM have out-performance in all experiments than the ensemble models. The better performance of the proposed model cause quantifying quantitative data.

References

1. Kim HJ, Hong SE, Cha KJ (2020) seq2vec: analyzing sequential data using multi-rank embedding vectors. Electron Commerce Res Appl 43:101003. ISSN 1567-4223, https://doi.org/10.1016/j.elerap.2020.101003, https://www.sciencedirect.com/science/article/pii/S1567422320300806
2. Russac Y, Caelen O, He-Guelton L (2018) Embeddings of categorical variables for sequential data in fraud context. In: International conference on advanced machine learning technologies and applications (AMLTA2018). Cham, pp 542–552
3. Guo C, Berkhahn F (2016) Entity embeddings of categorical variables. arXiv:1604.06737
4. Daly A, Dekker T, Hess S (2016) Dummy coding vs effects coding for categorical variables: Clarifications and extensions. J Choice Modell 21:36–41. ISSN 1755-5345, https://doi.org/10.1016/j.jocm.2016.09.005, http://www.sciencedirect.com/science/article/pii/S1755534516300781

5. Oleghe O, Salonitis K (2015) Improving the efficacy of the lean index through the quantification of qualitative lean metrics. Proc CIRP 37:42–47. CIRPe 2015 - Understanding the life cycle implications of manufacturing. ISSN 2212-8271, https://doi.org/10.1016/j.procir.2015.08.079
6. Dua D, Graff C (2017) UCI machine learning repository. Institution University of California, Irvine, School of Information and Computer Sciences http://archive.ics.uci.edu/ml
7. Luengo J, Shim S-O, Alshomrani S, Altalhi A, Herrera F (2018) CNC-NOS: Class noise cleaning by ensemble filtering and noise scoring. Knowl-Based Syst 140:27–49. ISSN 0950-7051
8. Bequé A, Lessmann S (2017) Extreme learning machines for credit scoring: an empirical evaluation. Expert Syst Appl 86. https://doi.org/10.1016/j.eswa.2017.05.050

Chapter 4
Deep Learning Model for Product Classification

4.1 Hierarchical Text Classification Problem and Applications

The text classification problem is one of the common machine learning problems, where an input object can be assigned to one or more output labels. However, in some cases, data labels can be organized under a hierarchical structure called hierarchical text classification (HTC). In this problem, assuming an object is defined as belonging to a certain parent label, it can only be in a corresponding sublabel set. Depending on each different task, we need to fully define the labels in the parent class or just define the labels in the subclass. The hierarchy can be represented as a tree or as a directional graph that has no cycle. For example, consider the case where we want to catalog a products that is described as "Computer Macbook Pro 2021". Then we can classify the product under the category "Computer", "Laptop", "Macbook". Dividing products into hierarchical categories helps to classify goods so that they can be searched easily.

4.1.1 HTC Problem in Product Classification Based on Its Description

Currently, with the explosion of technology 4.0, e-commerce models gradually become more popular to replace conventional wholesale and retail firms. The product category is considered the backbone of every e-commerce platform, but it can be a difficult task for e-commerce managers to ensure that all products are classified under the correct category. The set of categories and products available is often huge, constantly changing, and new products to be added daily.

© The Author(s), under exclusive license to Springer Nature Singapore Pte Ltd. 2022
N. T. N. Anh et al., *Artificial Intelligence for Automated Pricing Based on Product Descriptions*, SpringerBriefs in Computational Intelligence,
https://doi.org/10.1007/978-981-16-4702-4_4

To improve the product classification process, the author proposes to develop a system of machine learning that can predict which category is best suited for a given product, to make the whole process easier, faster and fewer errors. The main problem is a hierarchical classification problem. Solving this problem faces some of the following major challenges:

- **The number of text labels is huge:** Machine learning models typically well predict between a limited number of labels about tens to hundreds of labels, but in e-commerce there are often hundreds or thousands of categories required. To train robust models for these cases, an exceptionally large amount of training data is needed.
- **Product category information is diverse and imbalanced:** A product may have a detailed set of attribute data, which is completely missing in another product (e.g. color, size or material of the T-shirt compared to the expiration date, fat content or volume of milk). Taking all available product variables into account will result in a lot of missing value, which makes it much more difficult to train the model.
- **Each store has its own category structure:** Each e-commerce store typically has different category structures. A product in category A in one store but in category B in another, depending on the store arrangement. So converting products back and forth between e-commerce sites will take a lot of time for administrators when the number of products as well as categories is very large.

4.1.2   BERT Model for Text Classification

BERT stands for Bidirectional Encoder Representations from Transformers, a new architecture for the language representation problem, published by Google in early November 2018. Unlike recent language representation modes, BERT is designed to pretrain deep bidirectional representations from unlabeled text by jointly conditioning on both left and right context in all layers. As a result, the pre-trained BERT model can be finetuned with just one additional output layer to create state-of-the-art models for a wide range of tasks, such as question answering and language inference, without substantial task specific architecture modifications. In this sentence classification model, the only output value used for the decoding class is the output value corresponding to the special character [CLS] at the beginning of each input string.

This character information represents the entire content of the sentence, so it has a key meaning in the classification model. Call $c_s \in R^m$ where m is the number of dimensions of the output vector of the encoding class corresponding to the character [CLS], we have the probability of the output value of the model as follows:

$$P = \text{softmax}(Wc_s + b) \tag{4.1}$$

4.1.3   BERT Model with Multi-layer Loss Function

In the case the output of data depends on each other in a hierarchical tree structure. In the first level it is a common text classification model, with the number of output being the number of labels at that level. However, considering the second level, we can see that the number of labels at this level will increase significantly compared to the first level and at the next level the number of labels will increase possibly exponentially. The number of labels the model has to learn will be huge and the learning may not be as good. This can be understood as a person with few tasks can solve each task better than one person having to solve many tasks. As far as this work breakdown is in mind, we can train multiple models for each subset of each node in the tree. For example, in the first level with c_1 labels we will use a classification model at level 1 with the output label number c_1. Considering level 2 with the number of labels c_2, we will group the output labels at level 2 into $c1$ group by its respective parent label value. So we can use c_1 to model the classification at the second level. Likewise, the third level will use c_2 the classification model. Thus, considering the kth level, we will need to use $1 + c_1 + c_2 + \cdots + c_{k-1}$ model. With such a large number of models, it is difficult to apply in practice because a large number of models will require larger computing equipment, faster processors. To overcome the above two problems, we can use $1 + c_1 + c_2 + \cdots + c_{k-1}$ model with a single encoder, although this may not be possible. Performance as using separate encoders but with a significant reduction in storage space and computation time. In order to be able to reuse such encoder, the author to export a training and prediction method as follows (Fig. 4.1).

Assume that the data label is made up of a tree with k levels with label set classes $C_1, C_2, \ldots, C_k$ corresponding to each level. The ith level has N_i labels, when it is determined that the input of label c_j is at level C_i, it can only belong to labels that are children of label c_j at level $i + 1$. Thanks to that constraint, we can reduce the risk of misclassifying other labels of the model. Instead of using 1 fully connected layer as in the normal classification model, we will use k fully connected layers for

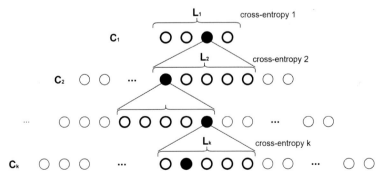

Fig. 4.1 Multi-layer loss function

each level. The full class at the ith level will have the output number of label C_i. The math formula is shown as follows:

$$P_1 = \text{softmax}(W_1 c_{cls} + b_1) \tag{4.2}$$
$$P_2 = \text{softmax}(W_2 c_{cls} + b_2) \tag{4.3}$$
$$\cdots \tag{4.4}$$
$$P_k = \text{softmax}(W_k c_{cls} + b_k) \tag{4.5}$$

The value P_i will be the softmax output of the labels at level i, W_i, b_i are the weighted values of the fully connected layer for the ith level.

While training the model, we know the possible label set at each level of data, so when calculating the value of the loss function, we only calculate the cross-entropy value on those labels.

$$\text{loss}_1 = \text{cross-entropy}(L_1, P_1) \tag{4.6}$$
$$\text{loss}_2 = \text{cross-entropy}(L_2, P_2) \tag{4.7}$$
$$\cdots \tag{4.8}$$
$$\text{loss}_k = \text{cross-entropy}(L_k, P_k) \tag{4.9}$$

where L_i is the possible label set of data at level i. To be able to represent it as matrix operations when performing calculations. We will use a mask vector with the values 0 and 1. Considering at the ith level, the mask vector M_{L_i} of the label set L_i will have a dimension number of labels C_i, the vector value at position j will be 1 if $c_j \in L_i$ and 0 if $c_j \notin L_i$. By correspondingly multiplying each component of the softmax P_i vector and the M_{L_i} mask vector, we can ensure that the softmax value of labels not in the L_i set will always be zero. Therefore the predicted result will only be a certain label in L_i set. Also from multiplying by a constant 0, when calculating the derivative of the function, we can remove a large number of connections at the full connection layer, so it will reduce a large amount of calculation. Note that the quantity of L_i is significantly smaller than C_i, which is higher at level i.

The final loss function value will be calculated by loss $= \sum_i \text{loss}_i$

In the result prediction step, we will perform label prediction of the input data from level 1 to level k respectively. The label set of level 1 is all labels at level 1, so $L_1 = C_1$. The label set of the second level are the sublabels of the predicted result at the first level L_2. Repeat until the kth level we get the label of the data at all levels and ensure parent and child binding of the data label.

4.2   Training Data

Data used to evaluate the model is descriptive data about products crawled by the author from the website https://tiki.vn. This is one of the largest electronics sales websites in Vietnam with a large and diverse product range including electronics, electronics to household appliances, everyday items, ... The crawler was written in python using the selenium and BeautifulSoup library.

Using some techniques on the browser we can get product-related information such as images, short descriptions, full description, ... of the product. Although all the information is of great significance used to identify the product type. However, we only focus on determining the type of product based on the description of that product, namely the information in the "short_description" field.

For each product we use the descriptive information as the input to the model, while the output or the model label is the category name of that product. For example, here the Macbook Pro 2020 M1 computer will be in the category "Laptop— Computers—Components" -> "Laptop" -> "Macbook". The list is organized in a multi-level format, so we will also build a set of labels in multiple levels. Specifically, Macbook Pro 2020 M1 computer will have label level 1 as "Laptop—Computers— Components", label level 2 is "Laptop", label level 3 is "Macbook". The data parameters given in this thesis are done by the author and the statistics in November 2020. The data after crawler included 11489 products with a maximum catalog depth of 6 levels, the number of labels per level.

Because the crawler data is abundant, we will filter out labels with quantity less than 20 and the remaining 103587 products with 5 levels as Table 4.1.

Because the number of labels at the $4th$ and $5th$ level are similar, in this assessment we only consider the 4th level of the label with the number of labels of each level being 14, 116, 549, 948.

To evaluate the model, we divide the data into 2 trainings and test with the number of 60000 and 43587 products, respectively. The number of products in each category is given in Table 4.2.

Because the number of labels and sub-categories is quite large (more than 900 sub-categories at level 4), we only mentioned the number of a few categories with high volume (Table 4.3).

Table 4.1 tNumber of tiki product categories that contain more than 20 products

Level	Number of labels
1	14
2	116
3	549
4	948
5	985

Table 4.2 Number of categories in train and test set

Category	Train	Test
Online department store	5732	4098
Car—Motorcycle—Bike	5539	4020
Electric appliances	5521	4136
Digital equipment—Digital accessories	5501	3970
Sports—Picnic	5463	3823
Camera camcorder	5426	3908
Tiki bookstore	5219	3844
Beauty—Health	4997	3795
Toy—Mother and Children	4615	3386
House—Life	4494	3216
Electronic refrigeration	4045	2916
Voucher—Service	2831	2069
Laptop—Computer—Accessories	360	235
Phone—Tablet	257	171

4.3   Evaluation Metrics

To evaluate the results of the model we will use accuracy and F1-score.

Accuracy
The correct scale of the model is calculated by the amount of correctly predicted data on the total number of data.

$$Acc = \frac{\text{Number of true predictions}}{\text{Total data}} \tag{4.10}$$

F1-score
Assuming that on label A, we denote TP (True Positive) as the number of entities of type A that are correctly classified, FP (False Positive) is the number of entities of type A that are classified incorrectly, FN (False Negative) is an entity number that is not of type A has been misclassified in it. From there, we determine the evaluation indicators by the following formula.

- Precision: is the ratio of the number of correctly classified entities among those that have been classified as type A.

$$\text{precision} = \frac{TP}{TP + FP}$$

Table 4.3 Structure and number of labels in train and test sets

Category	Train	Test
Online department store	5732	4098
I + Food	1176	857
I I + Dry food	960	701
I I I + Instant food	408	277
I I I + Dried fruit seeds	393	322
I I I + ...	159	102
Car Motorcycle bike	5539	4020
I + Accessories car care	5244	3771
I I + Car accessory	1689	1179
I I I + Home furniture	1069	771
I I I + Car exterior accessories	611	402
I I I + ...	9	6
Electric Appliances	5521	4136
I + Kitchen utensils	3267	2406
I I + Electric cookers of all kinds	822	652
I I I + Electric cooker	349	308
I I I + Frying pan	199	128
I I I + ...	274	216
Camera Camcorder	5426	3908
I + Accessories for cameras, camcorders	1959	1401
I I + Accessories for cameras, camcorders other	756	532
I I + Battery charger for cameras and camcorders	283	190
I + Surveillance camera	1736	1229
I I + Camera IP	1114	777
I I + Surveillance camera Analog	245	154
Digital Equipment Digital Accessories	5501	3970
I + Phone and tablet accessories	2795	1975
I I + Leather back cover	1072	774
I I I + Iphone leather back cover	329	247
I I I + Samsung leather back cover	312	217
I I I + ...	431	310
I + Audio equipment and accessories	1203	909
I I + Music speakers	427	333
I I I + Bluetooth speaker	257	179
I I I + ...	170	154

- Recall: The ratio of the number of entities that are correctly classified in type A among those that are actually of type A.

$$recall = \frac{TP}{TP + FN}$$

- F1-score: is the equilibrium representation of Precision and Recall.

$$F1\text{-}score = \frac{2 * precision * recall}{precision + recall}$$

In the case of evaluating a model with multiple labels, we often use additional methods of calculating the average rating of micro and macro.

- Micro average: This method will calculate the total contribution of the labels to the overall result. Value example

$$precision_{micro} = \frac{\sum TP}{\sum TP + \sum FP}$$

- Micro average: This method will calculate the total contribution of the labels to the overall result. Value example

$$precision_{macro} = \frac{1}{n} \sum \frac{TP}{TP + FP}$$

These two methods will give relatively different values when our data is unbalanced. The micro value will give us the evaluation results based on the distribution of the data label, while the macro value will give us an even evaluation of the labels.

4.4 Experiment

The experiment model uses the $BERT_{BASE}$ pre-training model with multi-lingual data combined with the hierarchical loss function with the number of levels of 4. In addition, the author also trains the feature base model using Tf-Idf combined with logistic model and classification model using $BERT_{BASE}$. The parameters of the models are as follows:

- TF-IDF + Logistic: Using the Tf-Idf method with n-grams from 1 to 3, remove words that appear 5 times less and words that appear more than 50% of the data. Classification model using logistic linear regression neural networks in the Sklearn library.
- BERT: Use the pre-training model $BERT_{BASE}$ with L = 12, H = 768, A = 12 with a total of 110 million parameters. Trained on multilingual data tissue. The pre-

Table 4.4 Model evaluation with category label level 1

Model	Level 1 (14 labels)			
	Accuracy	Precision (micro/macro)	Recall (micro/macro)	f1-score (micro/macro)
Tf-idf + logistic	86.8	86.8/88.9	86.8/81.5	86.8/83.4
$BERT_{BASE}$	86.5	86.5/86.6	86.5/85.9	86.5/86.2
$BERT_{BASE}$ + hierarchical loss	90.9	90.9/91.6	90.9/90.5	90.9/91.0

Table 4.5 Model evaluation with category label level 2

Model	Level 2 (116 labels)			
	Accuracy	Precision (micro/macro)	Recall (micro/macro)	f1-score (micro/macro)
Tf-idf + logistic	80.0	80.0/79.8	80.0/55.9	80.0/61.8
$BERT_{BASE}$	81.9	81.9/75.7	81.9/73.3	81.9/73.6
$BERT_{BASE}$ + hierarchical loss	86.1	86.1/79.2	86.1/75.1	86.1/76.6

training model weights are downloaded from https://huggingface.co/bert-base-multilingual-cased

- BERT + hierarchical loss: Using the $BERT_{BASE}$ pre-training model combined with hierarchical loss function with 4 levels and the number of output neurons at levels 14, 116, 549 and 948.

All 3 models are trained with a data set containing 60000 commodities described above. The training time of the Logistic classification model is about 30 min on the personal computer. BERT model and BERT model + hierarchical loss train 50 epoch on server with P100 GPU for about 6 h.

For Logistic and $BERT_{BASE}$ classification models. Although the results at levels 1, 2, and 3 are likely to be higher when we train the model with each level. However, since the end goal is the predicted result at level 4, we will predict the label level 4 of the product and then redefine the labels level 1, 2, 3 for evaluation.

Since the data has a very large number of labels and is heavily skewed on some popular product labels, we will use both the micro and macro numbers for a more complete view. The test results on the test set with the number of 43587 products are presented in Tables 4.4, 4.5, 4.6, 4.7, 4.8.

Based on the evaluation tables, we can see that the results of the model are significantly reduced when the number of data labels increases at all levels. At the first level with 14 data labels, both the featue-base and BERT models give relatively high results. But when classified to level 2, 3 with increased number of labels, the feature-base model proved much worse than BERT (at the 4th level the $f1_{macro}$ value set only 39.9% compared to 58.2% of the model. BERT and 60.0% on BERT + hierar-

Table 4.6 Model evaluation with category label level 3

Model	Level 3 (549 labels)			
	Accuracy	Precision (micro/ macro)	Recall (micro/ macro)	f1-score (micro/ macro)
Tf-idf + logistic	67.5	67.5/62.5	67.5/41.5	67.5/45.8
$BERT_{BASE}$	73.3	73.3/66.4	73.3/62.6	73.3/63.1
$BERT_{BASE}$ + hierarchical loss	75.9	75.9/68.3	75.9/65.1	75.9/65.8

Table 4.7 Model evaluation with category label level 4

Model	Level 4 (948 labels)			
	Accuracy	Precision (micro/ macro)	Recall (micro/ macro)	f1-score (micro/ macro)
Tf-idf + logistic	59.1	59.1/58.1	59.1/36.5	59.1/39.9
$BERT_{BASE}$	67.6	67.6/62.1	67.6/57.8	67.6/58.2
$BERT_{BASE}$ + hierarchical loss	69.1	69.1/62.3	69.1/59.7	69.1/60.0

Table 4.8 Top k accuracy assessment of the model at the 4th level

Model	Top 1 accuracy	Top 3 accuracy	Top 5 accuracy
$BERT_{BASE}$	67.6	72.1	72.9
$BERT_{BASE}$ + hierarchical loss	69.1	74.3	75.4

chical loss model). The results of the average macro rating are relatively inferior to the micro mean because there are several labels with large amounts of data classified with high accuracy that contribute some of the results end. The use of the hierarchical loss function improves the model's results because the label structure of the data can be taken advantage of. On all evaluation indicators, the BERT model with decentralized loss function gives more outstanding results, for levels 1, 2, 3 loss functions help improve the model up to 4%. In the 4th level (948 labels) it improves 1.5% accuracy, 1.5% $f1_{micro}$ value and 1.8% $f1_{macro}$ value). When considering accuracy of top 3 and 5 labels at level 4, BERT + hierarchical loss model gives 5–6% higher result than top 1 and asymptotic to accuracy at 3rd level.

Once again shows us the great strength of the pre-trained BERT model in the text classification problem compared to the traditional feature selection models such as TF-IDF. The use of a loss function with a hierarchical architecture improves the results of the BERT model with almost equivalent computation costs.

References

1. Samuel AL (1959) Some studies in machine learning using the game of checkers. IBM J Res Dev 3(3):210–229
2. Wilson E, Tufts DW (1994) Multilayer perceptron design algorithm. In: Proceedings of IEEE workshop on neural networks for signal processing, pp 61–68
3. Zhang Y, Jin R, Zhou Z-H (2010) Understanding bag-of-words model: a statistical framework. Int J Mach Learn Cybern 1:43–52
4. Goodman J (2001) A bit of progress in language modeling, CoRR, vol. cs.CL/0108005
5. Kuhn R, De Mori R (1990) A cache-based natural language model for speech recognition. IEEE Trans Pattern Anal Mach Intell 12(6):570–583
6. Andreas J, Vlachos A, Clark S (2013) Semantic parsing as machine translation. In: Proceedings of the 51st annual meeting of the association for computational linguistics (Vol 2: Short Papers), (Sofia, Bulgaria). Association for Computational Linguistics, pp 47–52
7. Lebret R, Collobert R (2014) Word embeddings through hellinger PCA. In: Proceedings of the 14th conference of the European chapter of the association for computational linguistics, (Gothenburg, Sweden). Association for Computational Linguistics, pp 482–490
8. Kouw WM (2018) An introduction to domain adaptation and transfer learning, CoRR, vol. abs/1812.11806
9. Mikolov T, Chen K, Corrado G, Dean J (2013) Efficient estimation of word representations in vector space
10. Pennington J, Socher R, Manning C (2014) GloVe: global vectors for word representation. In: Proceedings of the 2014 conference on empirical methods in natural language processing (EMNLP), (Doha, Qatar). Association for Computational Linguistics, pp 1532–1543
11. Bojanowski P, Grave E, Joulin A, Mikolov T (2017) Enriching word vectors with subword information. Trans Assoc Comput Linguist 5:135–146
12. Devlin J, Chang M-W, Lee K, Toutanova K (2018) BERT: pre-training of deep bidirectional transformers for language understanding. arXiv:1810.04805
13. Vaswani A, Shazeer N, Parmar N, Uszkoreit J, Jones L, Gomez AN, Kaiser L, Polosukhin I (2017) Attention is all you need CoRR, vol. abs/1706.03762
14. Dreiseitl S, Ohno-Machado L (2002) Logistic regression and artificial neural network classification models: a methodology review. J Biomed Inf 35(5):352–359

Chapter 5
Product Feature Extraction from the Descriptions

5.1 Semi-automatic Labeling Model Based on Doccano Framework

5.1.1 Labeling Car Description Data Problem and Overall Labeling Model

Data provides information about accessories, equipment, ... related to cars. The purpose of the problem is to label data such as brand, model,... from the automobile description. For example, the record has an item description:

> "*Xe ô tô con hiệu: MERCEDES BENZ G63 đã qua sử dụng, tay lái bên trái, SX: 2013. Cam kết không đục sửa số khung số máy, chi tiết như bảng kê đính kèm.*"

Then we need to identify the brand: MERCEDES BENZ and the model: G63

The overall labeling model combining expert and machine learning consists of two main parts: manual and automatic labeling as Fig. 5.1.

5.1.2 Expert-Based Labeling Process

Doccano system is an open source web tool that allows text labels to create training datasets. Doccano can connect many users with the same label: the requester uploads data to the tool, assigns the work to the user through the tool, the user can work and submit the job at the tool. It can also evaluate work progress. Doccano includes 3 types of projects, each of which supports the labeling of several problems:

- The first type is Text classification which supports labeling for text classification problems.

N. T. N. Anh et al., *Artificial Intelligence for Automated Pricing Based on Product Descriptions*, SpringerBriefs in Computational Intelligence, https://doi.org/10.1007/978-981-16-4702-4_5

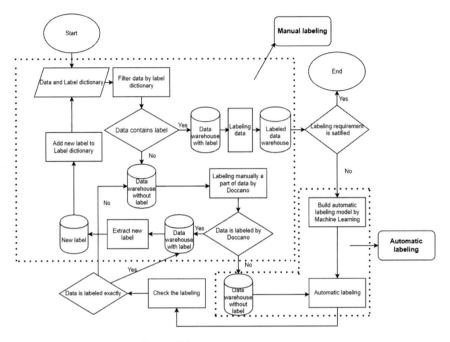

Fig. 5.1 Semi-automatic labeling model

- The second type is Sequence Labeling which supports labeling for Information extraction problems.
- The third type is Sequence to Sequence, which supports tagging for Text summarization / Machine translation problems.

Here are the steps in the expert-based labeling process:

Step 1: We define the initial input consisting of a data set describing the car goods and the label dictionary. The label dictionary may not contain all labels, so it is necessary to add and update the label dictionary.

Step 2: We filter the data describing automobile goods (the value in column "TEN_HANG") initially according to the label dictionary. From the data, we use the label in the label dictionary to search (search) in records of automobile description data. Then, our data set after filtering is divided into 2 parts: labeled data and unlabeled data. If the amount of labeled data is large enough (100% of the original data or not but enough to build the automatic labeling machine learning model) go to Step 4, otherwise go to Step 3.

Step 3: For unlabeled data, we proceed with manual labeling based on the Doccano framework, we can label it with an arbitrary number of people available. After labeling with Doccano, this data set can be divided into 2 parts: one part is the new labeled data with Doccano, we use this section to get the newly added labels and update label dictionary. We obtain a new label dictionary, the rest is unlabeled data. If there exists

Table 5.1 First brand labeling based on expert

	label	start	end	Description
0	BMW	30	33	ÔTÔCON,7CHỖ,MỚI100%,5CỬA,HIỆU BMW 218I GRAN TO…
1	AUDI	18	22	Xe Ô Tô con 5 chỗ AUDI A6 1.8 TFSI, DT 1798CC,…
2	BMW	30	33	ÔTÔCON,5CHỖ,MỚI100%,4CỬA,HIỆU BMW 320I LIMOUSI…
3	LAND ROVER	21	31	Ô tô con 5 chỗ, hiệu LAND ROVER RANGE ROVER AU…
4	Chevrolet	12	21	Xe ô tô con Chevrolet Trailblazer 2.5L 4x2 MT …
…	…	…	…	
3042	TOYOTA	3	9	XE TOYOTA LANDCRUISER, SỐ KHUNG: HDJ810020380,…
3043	LAND ROVER	29	39	Xe ôtô đã qua sử dụng, hiệu: LAND ROVER RANGE …
3044	MINI	29	33	ÔTÔCON,5CHỖ,MỚI100%,5CỬA,HIỆUMINI COOPER S 5-T…
3045	MERCEDES-BENZ	17	30	Xe ô tô con hiệu MERCEDES-BENZ V250 AVANTGARDE…
3046	LAND ROVER	3	13	XE LAND ROVER P4 80, SỐ KHUNG: 645001304, SỐ M…

3047 rows × 4 columns

Table 5.2 First model labeling based on expert

	Description	label	start	end
0	ÔTÔCON,7CHỖ,MỚI100%,5CỬA,HIỆU BMW 218I GRAN TO…	218I	33	38
1	Xe ô tô 2 cầu hiệu Toyota Fortuner, mới 100%, …	Fortuner	26	34
2	Xe ÔTô con 5 chỗ AUDI Q5 Design 2.0 TFSI quatt…	Q5	21	24
3	Xe điện dùng để chở người, 8 chỗ, số khung/máy…	EG2068K	176	183
4	Xe ÔTô con 7 chỗ Audi Q7 45 TFSI quattro,DT 19…	Q7	21	24
…	…	…	…	…
3152	ôtô 5chỗ Mercedes-Benz GLE400 4MATIC (GLE400 4…	GLE400	22	29
3153	Xe ô tô con mới 100% hiệu LEXUS RX 350 AWD, ta…	RX 350	31	38
3154	ÔTÔCON,5CHỖ,MỚI100%,4CỬA,HIỆU BMW 320I LIMOUSI…	320I	33	38
3155	Ô tô 5 chỗ,4 cửa,hiệu Porsche Cayenne,xe SUV,2…	Cayenne	30	37
3156	Xe ô tô TOYOTA YARIS 1.3 Thể tích thực 1299cm3…	YARIS	15	20

3157 rows × 4 columns

a portion of the unlabeled data we can use the automatic labeling machine learning model to label it. Go back to Step 2.

Step 4: From the labeled data, we will label the data: Get the label corresponding to the data, proceed to build a new data set including item description, label name, position of the first character of label, the position of the last character of the label. Results of expert-based labeling experiment:

- After filtering data by label one time, there were 3047 labeled data records and 381 unlabeled data records, of the 381 unlabeled data records, 150 records were used for labeling based on Doccano (As a result, there were 109 labeled data added to the labeled data, the remaining 41 were unlabeled and should be omitted), the remaining 231 unlabeled records were labeled machine-based.
- After filtering data by vehicle series 1 time, capturing 3157 labeled data records and 271 unlabeled data records, of 271 data records without label, use 125 records for labeling. Based on Doccano, the recorded recordings are marked based on machine learning.

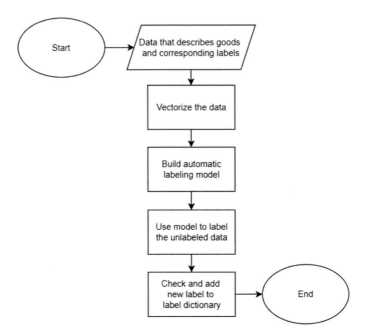

Fig. 5.2 Machine learning-based labeling process flowchart

{185: '@', 2: ' ', 3: ',', 4: 'n', 5: 'ø', 6: '1', 7: 'c', 8: 't', 9: 'h', 10: 'i', 11: 'S', 12: '2', 13: 'g', 14: 'u', 15: 'M', 16: 'A', 17: 'C', 18: 'a', 19: ':', 20: 'T', 21: '5', 22: '8', 23: 'e', 24: '4', 25: '9', 26: 'X', 27: 'm', 28: 'L', 29: 'N', 30: 'o', 31: 's', 32: '3', 33: 'D', 34: 'I', 35: 'õ', 36: 'E', 37: '7', 38: 'l', 39: '6', 40: 'd', 41: 'd', 42: 'á', 43: 'R', 44: 'K', 45: 'U', 46: 'x', 47: 'B', 48: 'O', 49: '.', 50: 'k', 51: 'r', 52: 'H', 53: 'V', 54: 'G', 55: 'y', 56: 'Y', 57: 'ő', 58: 'W', 59: 'b', 60: 'õ', 61: 'ộ', 62: 'ử', 63: 'ạ', 64: '%', 65: 'P', 66: 'Z', 67: '-', 68: 'ệ', 69: 'ớ', 70: 'à', 71: 'ó', 72: ')', 73: 'F', 74: 'ự', 75: 'ả', 76: 'ề', 77: 'Á', 78: 'p', 79: 'ư', 80: 'í', 81: 'v', 82: 'ơ', 83: ';', 84: 'Ả', 85: 'ụ', 86: 'ậ', 87: '(', 88: 'Đ', 89: ')', 90: 'ầ', 91: 'q', 92: 'ế', 93: 'ể', 94: 'ò', 95: 'ạ', 96: '/', 97: 'Ở', 98: 'ẵ', 99: 'ạ', 100: 'ấ', 101: 'Ồ', 102: 'Ã', 103: 'ử', 104: 'Ệ', 105: 'Ọ', 106: 'À', 107: 'Ẽ', 108: 'ự', 109: 'Ờ', 110: 'ü', 111: 'ì', 11 2: 'z', 113: 'ẻ', 114: 'ẳ', 115: 'Ợ', 116: 'ủ', 117: 'ờ', 118: 'ằ', 119: '#', 120: 'ồ', 121: 'ở', 122: 'ỏ', 123: 'f', 124: 'w', 125: 'ỷ', 126: 'ị', 127: 'ổ', 128: '"', 129: 'ẫ', 130: '+', 131: 'ữ', 132: 'ỉ', 133: 'ẩ', 134: 'Ẩ', 135: 'Ỉ', 136: 'ợ', 137: 'Ẫ', 138: 'Ế', 139: 'ỗ', 140: 'Ẵ', 141: 'ỹ', 142: 'j', 143: 'ọ', 144: '=', 145: '?', 146: 'Ạ', 147: 'Ọ', 148: '_', 149: 'ỗ', 15 0: 'Ữ', 151: 'ẹ', 152: 'Ẳ', 153: 'ỵ', 154: 'ỡ', 155: 'ỳ', 156: 'Ý', 157: '\r', 158: '\n', 159: 'ứ', 160: '"', 161: 'Ị', 162: 'Ẩ', 163: 'ẹ', 164: 'ẽ', 165: 'ẻ', 166: 'Ì', 167: 'Ỵ', 168: 'Ỉ', 169: 'ố', 170: 'ạ', 171: 'ồ', 172: 'Ấ', 173: 'Ừ', 174: 'ỏ', 17 5: 'Ọ', 176: 'Ề', 177: 'Ể', 178: 'Ỉ', 179: 'ú', 180: 'Ồ', 181: 'ẳ', 182: 'ữ', 183: 'Đ', 184: '>', 186: '\t', 187: 'Ể', 0: ''}

Fig. 5.3 Illustration of dictionary of characters

5.1.3 Machine Learning-Based Labeling Process

In this section, we build an automatic labeling machine learning model for item description data.

At the vectorization step, we need to convert the data from text to digital to build an automatic labeling machine learning model by creating a dictionary of all characters appeared in the corresponding data and index:

To satisfy the input to the machine learning model (Bi-LSTM), the coded vector of the label is slightly modified, each element in the vector is divided into a vector of 2 elements, if an element is in the sequence. vector is 0 then the new vector is [1 0], if it is equal to 1 then the new vector is [0 1].

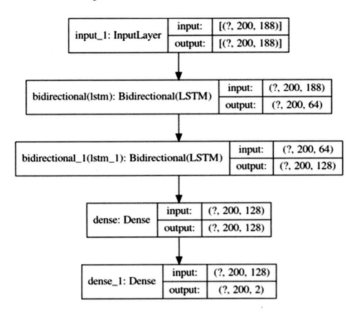

Fig. 5.4 Deep learning model architecture

The next step is to build a deep learning model to label the unlabeled data. The model architecture is depicted in Fig. 5.4.

After building a deep learning model, we use this model to automatically label the brand and model of the car product. The next step in the overall labeling model combining expert and machine learning is post-check to check that the records are correctly labeled, if correct, input the label, extract a new label, add a new label. add new labels to the label dictionary. If this is not correct, import the unlabeled data for labeling with Doccano.

After completing the branding and vehicle model for the data, we have the final data set:

- For labeling the brands: we get 3350 records that contain labels, and the rest don't contain labels, we ignore them.
- For labeling the models: we have 3281 labeled records, the remaining records do not contain labels we ignore.

Table 5.3 Labeling brand dataset

	label	start	end	Description
0	BMW	30	33	ÔTÔCON,4CHỖ,MỚI100%,2CỬA,HIỆU BMW 420I CABRIO,...
1	Mitsubishi	26	36	Xe ôtô con hatchback,hiệu Mitsubishi MIRAGE GL...
2	AUDI	18	22	Xe Ô Tô con 5 chỗ AUDI A6 1.8 TFSI, DT 1798CC,...
3	BMW	30	33	ÔTÔCON,7CHỖ,MỚI100%,5CỬA,HIỆU BMW 218I GRAN TO...
4	MERCEDES-BENZ	18	31	Xe ô tô con hiệu: MERCEDES-BENZ S550 đã qua sử...
...	...	...	...	...
3345	BMW	30	33	ÔTÔCON,7CHỖ,MỚI100%,5CỬA,HIỆU BMW 218I GRAN TO...
3346	TOYOTA	22	28	Xe ô tô con nhãn hiệu TOYOTA RUSH,kiểu xe SUV,...
3347	SUBARU	21	27	Ô tô con 5 chỗ; hiệu SUBARU OUTBACK 2.5i-S EYE...
3348	LAND ROVER	18	28	Xe ô tô con hiệu: LAND ROVER RANGE ROVER đã qu...
3349	Porsche	21	28	Ôtô 5 chỗ,4 cửa,hiệu Porsche Cayenne,xe SUV,2 ...

3350 rows × 4 columns

Table 5.4 Labeling model dataset

	Description	label	start	end
0	ÔTÔCON,5CHỖ,MỚI100%,5CỬA,HIỆUMINI COOPER S 5-T...	COOPER	34	40
1	Ô tô con 5 chỗ; hiệu SUBARU Forester 2.0i-S Ey...	Forester	28	36
2	ÔTÔCON,5CHỖ,MỚI100%,5CỬA,HIỆU BMW X1 SDRIVE18I...	X1	34	36
3	Ô tô con 5 chỗ, hiệu LEXUS RX350 F SPORT, SK: ...	RX350	26	32
4	Xe chở người 4 bánh có gắn động cơ điện loại 7...	LT-S8	138	143
...	...	...	...	...
3276	Xe ôtô con nhãn hiệu: KIA OPTIMA; Đã qua sử dụ...	OPTIMA	26	32
3277	ÔTÔCON,5CHỖ,MỚI100%,5CỬA,HIỆUBMW X2 SDRIVE20I,...	X2	33	35
3278	Xe ôtô đã qua sử dụng, hiệu: BMW X5, cam kết k...	X5	33	35
3279	Ôtô con7chỗ5cửa,Hiệu:Land Rover,model:Discover...	Discovery Sport HSE	38	57
3280	ÔTÔCON,5CHỖ,MỚI100%,4CỬA,HIỆU BMW 320I LIMOUSI...	320I	33	38

3281 rows × 4 columns

5.2   Extracting Features of Vehicles from the Descriptions

5.2.1   Data Description

There are 2 data sets: Trademark data set and vehicle model tag data set. A trademark data set includes 4 fields (item description, label value, the start position of the label in the item description, the end position of the label in the item description) and 3350 records, the vehicle flow tagging data consists of 4 fields similar to the branding data set and 3281 records (See Tables 5.5 and 5.6).

Table 5.5   Car brand dataset

	label	start	end	Description
0	BMW	30	33	ÔTÔCON,4CHỖ,MỚI100%,2CỬA,HIỆU BMW 420I CABRIO,...
1	Mitsubishi	26	36	Xe ôtô con hatchback,hiệu Mitsubishi MIRAGE GL...
2	AUDI	18	22	Xe Ô Tô con 5 chỗ AUDI A6 1.8 TFSI, DT 1798CC,...
3	BMW	30	33	ÔTÔCON,7CHỖ,MỚI100%,5CỬA,HIỆU BMW 218I GRAN TO...
4	MERCEDES-BENZ	18	31	Xe ô tô con hiệu: MERCEDES-BENZ S550 đã qua sử...
...	...	...	...	...
3345	BMW	30	33	ÔTÔCON,7CHỖ,MỚI100%,5CỬA,HIỆU BMW 218I GRAN TO...
3346	TOYOTA	22	28	Xe ô tô con nhãn hiệu TOYOTA RUSH,kiểu xe SUV,...
3347	SUBARU	21	27	Ô tô con 5 chỗ; hiệu SUBARU OUTBACK 2.5i-S EYE...
3348	LAND ROVER	18	28	Xe ô tô con hiệu: LAND ROVER RANGE ROVER đã qu...
3349	Porsche	21	28	Ôtô 5 chỗ,4 cửa,hiệu Porsche Cayenne,xe SUV,2 ...

3350 rows × 4 columns

Table 5.6   Car model dataset

	Description	label	start	end
0	ÔTÔCON,5CHỖ,MỚI100%,5CỬA,HIỆUMINI COOPER S 5-T...	COOPER	34	40
1	Ô tô con 5 chỗ; hiệu SUBARU Forester 2.0i-S Ey...	Forester	28	36
2	ÔTÔCON,5CHỖ,MỚI100%,5CỬA,HIỆU BMW X1 SDRIVE18I...	X1	34	36
3	Ô tô con 5 chỗ, hiệu LEXUS RX350 F SPORT, SK: ...	RX350	26	32
4	Xe chở người 4 bánh có gắn động cơ điện loại 7...	LT-S8	138	143
...	...	...	...	...
3276	Xe ôtô con nhãn hiệu: KIA OPTIMA; Đã qua sử dụ...	OPTIMA	26	32
3277	ÔTÔCON,5CHỖ,MỚI100%,5CỬA,HIỆUBMW X2 SDRIVE20I,...	X2	33	35
3278	Xe ôtô đã qua sử dụng, hiệu: BMW X5, cam kết k...	X5	33	35
3279	Ôtô con7chỗ5cửa,Hiệu:Land Rover,model:Discover...	Discovery Sport HSE	38	57
3280	ÔTÔCON,5CHỖ,MỚI100%,4CỬA,HIỆU BMW 320I LIMOUSI...	320I	33	38

3281 rows × 4 columns

5.2.2   Building Model

For each individual problem (brand and model feature extraction), the data are divided 80-20 for the training data set (train) and the test dataset. All data are pre-processed before being put into the model. The general architecture of the model is presented in Fig. 5.5.

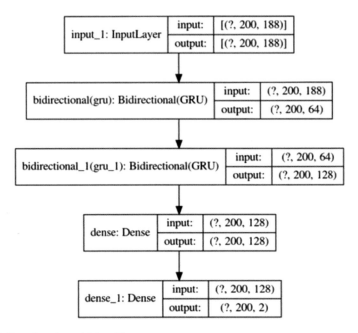

Fig. 5.5 Deep learning model architecture

5.2.3   Evaluating Model

Here, F1 and Accuracy points (accuracy) are used to evaluate the model. Formula for calculating point F1: Let set $\hat{Y}$ be the set of values predicted by the model, set Y is the set of values actually obtained from the label, then we have:

$$\text{precision} = \frac{|\hat{Y} \cap Y|}{|\hat{Y}|}$$

$$\text{Recall} = \frac{|\hat{Y} \cap Y|}{|Y|}$$

$$F1 = \frac{2 \times \text{precision} \times \text{recall}}{\text{precision} + \text{recall}}$$

The calculation formula for Accuracy is based on perfectly accurately predicted labels:

$$\text{Accuracy} = \frac{\text{Total number of labels is correctly predicted}}{\text{The total number of test set labels}}$$

The computational results are obtained:

- For car brands: Accuracy is 96.27% and F1 is 94.00%.
- For car models: Accuracy is 89.64% and F1 is 94.28%.

References

1. Samuel AL (1959) Some studies in machine learning using the game of checkers. IBM J Res Dev 3(3):210–229
2. Wilson E, Tufts DW (1994) Multilayer perceptron design algorithm. In: Proceedings of IEEE workshop on neural networks for signal processing, pp 61–68
3. Zhang Y, Jin R, Zhou Z-H (2010) Understanding bag-of-words model: a statistical framework. Int J Mach Learn Cybernet 1:43–52
4. Goodman J (2001) A bit of progress in language modeling, CoRR, vol. cs.CL/0108005
5. Kuhn R, De Mori R (1990) A cache-based natural language model for speech recognition. IEEE Trans Pattern Anal Mach Intell 12(6):570–583
6. J. Andreas, A. Vlachos, and S. Clark, "Semantic parsing as machine translation," in Proceedings of the 51st Annual Meeting of the Association for Computational Linguistics (Volume 2: Short Papers), (Sofia, Bulgaria), pp. 47–52, Association for Computational Linguistics, Aug. 2013
7. R. Lebret and R. Collobert, "Word embeddings through hellinger PCA," in Proceedings of the 14th Conference of the European Chapter of the Asso- ciation for Computational Linguistics, (Gothenburg, Sweden), pp. 482–490, Association for Computational Linguistics, Apr. 2014
8. W. M. Kouw, "An introduction to domain adaptation and transfer learning," CoRR, vol. abs/1812.11806, 2018
9. T. Mikolov, K. Chen, G. Corrado, and J. Dean, "Efficient estimation of word representations in vector space," 2013
10. J. Pennington, R. Socher, and C. Manning, "GloVe: Global vectors for word representation," in Proceedings of the 2014 Conference on Empirical Methods in Natural Language Processing (EMNLP), (Doha, Qatar), pp. 1532–1543, Association for Computational Linguistics, Oct. 2014
11. Bojanowski P, Grave E, Joulin A, Mikolov T (2017) Enriching word vectors with subword information. Transactions of the Association for Computational Linguistics 5:135–146
12. J. Devlin, M.-W. Chang, K. Lee, and K. Toutanova, "BERT: pre-training of deep bidirectional transformers for language understanding," arXiv preprint arXiv:1810.04805, 2018
13. A. Vaswani, N. Shazeer, N. Parmar, J. Uszkoreit, L. Jones, A. N. Gomez, L. Kaiser, and I. Polosukhin, "Attention is all you need," CoRR, vol. abs/1706.03762, 2017
14. Dreiseitl S, Ohno-Machado L (2002) Logistic regression and artificial neural network classification models: a methodology review. Journal of Biomedical Informatics 35(5):352–359

Glossary

True positive A true positive is an true in binary classification

False positive A false positive is an error in binary classification

Accuracy Accuracy is one metric for evaluating classification models. Informally, accuracy is the fraction of predictions our model got right.

Precision Write here the description of the glossary term. Write here the description of the glossary term. Write here the description of the glossary term.

Recall Write here the description of the glossary term. Write here the description of the glossary term. Write here the description of the glossary term.

F_1 The F1 score is the harmonic mean of the precision and recall.

Regression regression analysis is a set of statistical processes for estimating the relationships between a dependent variable (often called the 'outcome variable') and one or more independent variables (often called 'predictors', 'covariates', or 'features').

Random Forest Random forests or random decision forests are an ensemble learning method for classification, regression and other tasks that operate by constructing a multitude of decision trees at training time and outputting the class that is the mode of the classes (classification) or mean/average prediction (regression) of the individual trees.

XGBoost is an open-source software library which provides a gradient boosting.

Light GBM Light Gradient Boosting Machine, is a free and open source distributed gradient boosting framework for machine learning.

© The Author(s), under exclusive license to Springer Nature Singapore Pte Ltd. 2022 53
N. T. N. Anh et al., *Artificial Intelligence for Automated Pricing Based on Product Descriptions*, SpringerBriefs in Computational Intelligence,
https://doi.org/10.1007/978-981-16-4702-4

Printed in the United States
by Baker & Taylor Publisher Services